TECHNOLOGY AND SOCIETY

TECHNOLOGY AND SOCIETY

Advisory Editor
DANIEL J. BOORSTIN, author of
The Americans and Director of
The National Museum of History
and Technology, Smithsonian Institution

THE NEW EPOCH

AS DEVELOPED
BY THE MANUFACTURE
OF POWER

BY GEORGE S. MORISON

ARNO PRESS

A NEW YORK TIMES COMPANY

New York • 1972

Reprint Edition 1972 by Arno Press Inc.

Reprinted from a copy in The University
of Illinois Library

Technology and Society
ISBN for complete set: 0-405-04680-4
See last pages of this volume for titles.

Manufactured in the United States of America

Library of Congress Cataloging in Publication Data

Morison, George Shattuck, 1842-1903.
    The new epoch as developed by the manufacture of
power.

    (Technology and society)
    Reprint of the 1903 ed.
    1. Technology and civilization. I. Title.
II. Series.
CB478.M65 1972   901.9   72-5064
ISBN 0-405-04715-0

THE NEW EPOCH
AS DEVELOPED BY THE
MANUFACTURE OF
POWER

Geo. S. Morison

THE NEW EPOCH

AS DEVELOPED
BY THE MANUFACTURE
OF POWER

BY GEORGE S. MORISON

BOSTON AND NEW YORK
HOUGHTON, MIFFLIN AND COMPANY
The Riverside Press, Cambridge

NOTE

THE preface of this volume gives the reasons for its preparation in its present form and the date at which it was prepared. Although it was ready for the press in April, 1898, it was not then published. There were several causes for this, one of which was a natural modesty that made the author hesitate to offer to the public a work which was outside the field of his usual professional writings. Mr. Morison died July 1, 1903. The manuscript is now published as he left it. The reader should occasionally bear in mind that it was completed before the war with Spain occurred and before the nineteenth century had closed.

Sept. 23, 1903.

PREFACE

THE general idea which runs through this
little book was first suggested to the writer
while reading his classmate John Fiske's
work entitled "The Discovery of America,"
in which the series of ethnical periods among
prehistoric men is elaborated. The fact that
the world is now entering on a new epoch
of the same nature is the fundamental idea
of the writer. This idea was subsequently
developed in an address, delivered in June,
1895, as President of the American Society
of Civil Engineers, which he called "The
New Epoch and the Civil Engineer;" it
was further elaborated in an oration delivered before the society of the Phi Beta
Kappa at Cambridge, Massachusetts, in
1896, to which he gave the title of "The
New Epoch and the University;" the same

subject was in a measure completed by an
address delivered at the annual commence-
ment of the Rensselaer Polytechnic Institute at Troy, New York, in June, 1897,
which was entitled "The Civil Engineer
and the University." Meanwhile the idea
had been used in an article entitled "The
New Epoch and the Currency," which was
prepared as a campaign document but was
too late to be used for that purpose, and
was subsequently published in the "North
American Review." Various friends have
requested that these several papers be put
in more permanent form, and this volume
is the result of such requests.

It seemed necessary to recast the subject,
but in doing this little change has been
made in the language, which may account
for some declamatory expressions which are
more appropriate in an address than in a
book. Chapter I. was practically common
to all three addresses. Chapter II. is taken
principally from the article in the "North

American Review." Chapters III., IV., and
VIII. are substantially new. Chapter V. is
principally from the presidential address,
Chapter VI. from the Phi Beta Kappa ora-
tion, and Chapter VII. from the Rensselaer
Polytechnic Institute address.

G. S. M.

CHICAGO, April 16, 1898.

CONTENTS

THE NEW EPOCH

AS DEVELOPED BY THE MANUFAC-
TURE OF POWER

I

GENERAL CONDITIONS

STUDENTS of primitive society have divided
the early development of the human race
into ethnical epochs, representing various
conditions of savagery and barbarism, and
finally culminating in civilization; they recognize three periods of savagery, followed
by three periods of barbarism. In the lowest
epoch men were little superior to the animals by which they were surrounded. With
the use of fire the second period began.
With the invention of the bow and arrow,
the most primitive form of projectile, man
entered the third period. With pottery, and
all that it implies, he passed from savagery

to barbarism. The next advance came with the domestication of animals, which gave man another power besides his own physical strength. With the manufacture of iron the last of the barbarous periods was entered. By the invention of the written alphabet the primitive race was promoted from barbarism to civilization.

The use of fire first placed man in a condition very different from that of other animals, giving him a power the uses of which are even yet not fully developed. The domestication of animals was hardly less important, and although where animals suitable for domestication did not exist tribes were able to pass this period without them, their weakness was apparent when they came in contact with other races whose conditions were not so limited. Finally the invention of a written language made the work of one generation available for its successors and produced historical civilization.

The changes which mark the advances from period to period are all material improvements ; in every instance they are characterized by some distinctive physical device which has enabled man either to utilize his own strength better than before, or to increase his power by adding other animate or inanimate force. The race that passed from one period to another acquired resources which it had not before ; in the contests which characterized the life of the primitive man, the men of a lower period fell before those who had risen higher. But though the devices were of a purely material character, they gave opportunities for mental and moral improvement which alone made further advance possible, till finally the written alphabet resulted in that preservation of knowledge which has made the intellectual efforts of thirty centuries available for ourselves. With the dawn of civilization the ethnical periods have been considered closed ; subsequent growth has been the natural advance of civilization marked by the events which make written history.

But there is no reason why the epoch

which began with writing should be the last. It only needed a new capacity, radically unlike those which have gone before, to make an epoch in civilization as distinct as those in primitive society. Such new capacity has now been found ; another epoch has begun. Fire, animal strength, and written language have in turn advanced men and nations ; something like a new capacity was developed with the discovery of explosives and again in the invention of printing ; but the capacity of man has always been limited to his own individual strength and that of the men and animals which he could control. His capacity is no longer so limited ; man has now learned *to manufacture power*, and with the manufacture of power a new epoch began. These words are used advisedly ; creation, whether of substance or force, is not given to man ; manufacture is not creation, but to change inert matter from one form to another in such way as to generate power is to manufacture power, and this we can do.

Furthermore, not only does the manufacture of power mark a new epoch in development, but the change is greater than
any which preceded it ; greater in its influence on the world ; greater in the results
which are to come.

The manufacture of power means that
wherever needed we can now produce practically unlimited power ; whatever the measure of a single machine, that machine can
be used to make a greater one ; we are no
longer limited by animal units, confined by
locations of waterfalls, nor angered by the
uncertain power of wind. Power can be had
where it is needed and when it is needed.
The power generated in a modern steamship in a single voyage across the Atlantic
is more than enough to raise from the Nile
and set in place every stone of the great
Egyptian pyramid.

The new epoch differs from all preceding
epochs, in that while they represented successive periods of progress, different races
have existed simultaneously in every period

of advancement, whereas the new epoch must from its very nature soon become universal. The manufacture of power has given us the means of traversing the entire globe with a regularity and speed which brings all races together, and which must in time remove all differences in capacity. It brings people of all races into contact, and, by extending knowledge, ends the superstitions and mysteries which have had such influence in the past. It enables man while working in unhealthy districts to spend a portion of his time in places favorable to physical health and bodily vigor, and so may end the climatic degeneration of race, which has done so much in history. It is gradually breaking down national divisions, substituting the natural boundaries of convenient government for boundaries based on race and ignorance. It will finally make the human race a single great whole, working intelligently in ways and for ends which we cannot yet understand.

It is not too much to predict that when

the full effects of the manufacture of power are realized and the world has passed through the development which the next ten centuries will see, that the time when man began to manufacture power will be recognized as the division between the ancient and the modern, between ignorance and intelligence, between the national strife which may then be classed as barbarism and the new civilization, whatever that may then be called.

The new epoch has barely begun. No exact dates can be fixed. Epoch making is not a matter of a single invention ; it is the general result which follows. It was not the manufacture of the first earthen pot, but the general introduction of pottery which carried a prehistoric race from savagery to barbarism. It was not the invention of a few letters, but the general use of a written language which took the barbarian into civilization. It was not the invention of the first steam engine, but the general control of the manufacture of power which is now taking mankind into the new civilization. James Watt developed his first steam engine in 1769. The steam engine began to come into general use about the beginning of this century. The nineteenth century has seen the development of the manufacture of power by steam. The steam engine is still almost the sole representative of manufactured power, but there is no reason why this should continue. Electricity as a conveyer of power has been developed to an extent which may almost be classed with manufactured power. New forms of manufactured power may come at any time, but the introduction of new forms is a comparatively unimportant thing. The great advance came with the ability to manufacture power at all; the method is a secondary thing.

It is easy to understand that when the new epoch is fully developed all physical work may be dependent on inanimate power. It is easy to see that this means the concentration of enormous masses of power where power never could be had before;

that it means the subdivision of power into units of a minuteness hard to conceive ; that it means the unraveling of mysteries which have never been solved ; that it means the construction of works of a magnitude before which the greatest monuments of antiquity become insignificant. The fighting ship of to-day is a floating machine-shop, though its crew of mechanics are confined as completely as the chained rowers of a Roman galley. The battles of the future will not be fought by men or by horses ; the camels of the desert will never again confront the elephants of the jungle ; fortifications will be factories filled with power. It is easy to recognize that the discoveries already made may be slight in comparison with those which are to come. All this is a matter of physical possibility ; it is interesting to speculate upon ; it is foolish to prophesy about ; these achievements are too close at hand for us to waste time in guessing what they will be.

The substitution of inanimate manufac-

tured power for the animal power on which
our race was formerly dependent means a
separation of the force which does the work
from the intellect which directs it. The
power which we make and use is absolutely
without sense ; all this must come from the
human mind. The man who drives a horse
has little to do ; the horse finds the way and
does the work. But the driver of a motor
carriage has a senseless machine, and all
direction must come from him. Manufactured power demands intelligence to supply
the sense which the power lacks. The extreme logical development would be a condition where every kind of physical work is
performed by machines, while human effort
is reduced to design and care. Such a result will never be reached. So long as men
have bodies, the forces placed in those
bodies must be used, but the substitution of
manufactured power for human labor is a
promotion for man, whose value becomes
measured by skill in directing power and not
by muscular strength.

No changes have ever equaled those through which the world is passing now. The manufacture of power has an intellectual as well as a physical effect ; it has separated power from the mind which must manage it ; it calls for intelligent design and direction of the multitude of works which it has rendered possible ; it has equipped our generation with tools for study and investigation as well as for mechanical work. The new epoch will alter the relations between the professions, business, and trades ; it will readjust the duties of government and the relations of one government to another ; it will change our system of education. These changes will be considered in relation to business, to national interests, and to education. The larger subject of government and international relations can only be touched upon briefly, but the effect of the new epoch is very important, and its influence in shaping the duties of government must not be overlooked. The group of new professions which are now coming into life will be reviewed and their proper influence on work outside of professional lines explored. The general demands of education in the new epoch will be considered, and some suggestions will be made as to the best methods of meeting these demands.

II

THE manufacture of power has entirely changed all methods of communication. The railroad has replaced the stage coach; the steamship has supplanted the graceful sailing vessel; and the telegraph has supplemented the laggardly mail. All this has been the work of the engineer.

In the early part of the century, to travel at the rate of one hundred miles a day was so difficult a thing that few people were willing to undertake it; now eight hundred miles can be accomplished with little fatigue in twenty-four hours, and even this speed has been materially exceeded. The first President of the republic spent weeks going from his home in Virginia to take the oath of office in New York; to-day there is no part of our country, except Alaska, from which a citizen cannot reach the national capital in less than six days. If we consider simply the time taken to travel personally, neglecting the other methods of communication, in which much greater speed has been attained, every part of the country has now become nearer the most distant section than Boston was to Richmond one hundred years ago ; this is what the railroad has done.

Only sixty years ago the entire business on the ocean was done by sailing vessels. They had improved in size, speed, and model since Columbus crossed in his caravels, but the only power to propel them was still the uncertain action of the wind. A good eastward voyage across the Atlantic consumed twenty days, while the return voyage usually took at least twice that time. This was the quickest method of communication between America and all the rest of the world. It was only under very exceptional conditions that any one could leave an American port and reach some of the nearer ports in

Europe and return in as short a time as
two months, and this was true not only of
men and women, but of mails and every
kind of communication. To-day it is possible to cross the Atlantic and return in less
than two weeks. The usual time occupied
by passenger steamers in a round voyage,
including the time which they wait in port
on both sides the Atlantic, and including
voyages to as distant ports as Bremen and
Hamburg, is now four weeks, while the
steamers of one of the leading steamship
lines are making a round voyage every three
weeks. The entire circuit of the globe can
be accomplished in less time than was commonly necessary in the early years of the
century for a hurried trip to a near European port and return.

The mails follow the same course as passengers ; they are taken on the fastest ships
and the fastest railroad trains, but they
have the advantage that they are transferred
from train to train and from train to ship
at the quickest speed, and are free from the

delays of inspection and sometimes of quarantine which annoy passengers. The cost of sending a letter to the most distant part of the globe is only one fifth as much as the cost of sending it from one state to another in the early days of the republic.

Great as these changes are, from a business point of view the effect of the telegraph is still greater. It has been able to annihilate time in communication between different towns, states, and countries; there is no important business centre in any part of the globe to which a message cannot be sent and a reply received within twenty-four hours, the delays generally being due not to the time occupied in transmission or delivery, but to the fact that the rotation of the earth and the different times of light and darkness make the business day in our continent simultaneous with the silent, dark hours of sleep in populous Asia.

The result of this quick communication has been an absolute change in all methods of doing business. Formerly each community was a unit by itself ; it handled its own business; its people seldom went outside of its own limit; and such luxuries as were obtained from a distance were brought at long intervals either by their owners or by people specially charged with handling them. The food, dress, and all the habits of each community were dependent on its immediate surroundings. If the soil and climate were adapted to wheat, the white bread of the present day was a common food, but in other regions where corn and rye grew better, wheat bread was seldom seen. Homespun garments were the usual wear ; the luxuries of imported fruits were unknown. The merchant in a seaport sent his ship to sea in charge of a supercargo whom he recognized as a capable business man, loaded with a cargo which that officer was to sell, and with the proceeds of which he was to make purchases for the return voyage. The chances were that from the time the ship left till it returned nothing would be heard of it. The capital invested

in the cargo was absolutely locked up, and
the merchant knew little or nothing of the
success of his venture until the ship, return-
ing from a voyage of perhaps a year, was
sighted off the harbor. It was in those days
that people talked of what they would do
when their ship came in. To-day any such
voyage is absolutely impossible. A commerce, of a tonnage before which that of
the old supercargo days becomes insignificant, is done by regular lines of steamers
sailing to almost every part of the world, on
regular schedules, from which they seldom
vary more than one or two days, and whose
arrival at every port is known at every
other port at which they touch, within
twenty-four hours of the time they reach
there. The sailing fleet of to-day no longer
goes out with ventures under the charge of
supercargoes, but the ship takes a load of
goods for which freight is paid, and which
is consigned to some merchant in the port
to which it sails, where the ship again loads
for another port; the captain sails the ship,

which is simply a means of transportation;
the business is handled by merchants in the
several ports, and every one of these merchants has the means of knowing what
goods are in demand, and what prices they
will bring in every other port in the world.
Commercially the whole earth has already
become a single unit, in every part of which
business can be directed from every other
part.

This is what the manufacture of power
has accomplished by its improvement in
methods of transportation and communication alone. No nation can live by itself; it
is a part of the whole commercial world; it
can be politically independent of other
countries, but it cannot be independent of
the general laws which govern trade and
commerce. Even if it desired to do so, it
would not be allowed to keep clear of the
developments of the nineteenth century.
The very system of commercial credits
which played so important a part in all
international commerce thirty years ago is

now passing away ; regular steamship lines
and telegraph cables have changed the practice ; the commercial standing of every firm
can be known everywhere, and the aid of
the few great houses is no longer needed ;
an event which would impair the standing
of any important business house in an in-
land American city may be felt in eastern
Asia.

The manufacture of power has effected at
least an equal change in all varieties of manufactures. Homespun garments have disap-
peared, complicated machinery directed by
few hands produces nearly every manufactured product which is essential to man. The
spinning wheels, which were a necessary part
of our great-grandmother's households, are
now kept only as curiosities in bric-a-brac
decorations, while the housewife who could
weave even her own supply of linen would
be very hard to find.

It is not necessary to go into the details
of this change. The result has been accomplished by substituting for the muscle of

arm and leg the power which the steam engine or the turbine wheel provides, power which is fed on coal at five cents a bushel instead of power fed on wheat at eighty cents, or which is furnished free in the slopes of rivers.

This is not merely the case with cloths and fabrics; it is true of every kind of manufacture and every kind of work. Power has so reduced the price of all the little articles which enter into our daily life that no one can afford to make them himself. In the older parts of the country there are old barns the frames of which were so made that the boards could be fastened in without nails; when they were built the price of nails was more than ten times what it is now, while the value of a workman's time was only about one third its present value.

The development of iron manufacture shows what has been done. Only twenty years ago nothing typified the strain of human labor more than the row of furnaces in which the puddlers, by muscular effort and in glaring heat, slowly drew together the particles of soft metal into the spongy puddle-ball from which wrought iron was forged and rolled. To-day the Bessemer converter and the open-hearth furnace have spoken the doom of wrought iron, which is disappearing before the less costly steel, and there is nothing more striking about a great steel plant than the absence of men. Ingots, blooms, billets, and finished product seem to make themselves, while the few men who stand around appear to police the machinery rather than to labor.

It was common in country districts in the early part of the century to find families who bought almost nothing ; their food was all raised on their farms, while the clothing and all the household linen was of wool or flax raised at home, spun and woven by the mothers and daughters of the family. To-day it is still possible to find places where people live well and where all the food is raised at home, but in the whole United States it is doubtful whether a hundred families can be found who do not buy their clothing and most of their household goods.

The results may be very briefly stated. Cheap manufactured power does the work; a few intelligent hands direct the machinery; a day's labor expended in other ways will buy ten times as much cloth or any other manufactured article as the good hands of our capable grandparents could make in a day.

Perhaps the manufacture of power has had less influence in agriculture than in anything else, but even here the effect has been enormous. In the first place, it has filled the world with a great variety of agricultural machinery, which has reduced the labors of the farmer to a small fraction of what they formerly were, till the same man can accomplish with these tools four or five times as much as he could do without them. But there is another remarkable change. The crops of every part of the world have become available for the rest of the world; a failure of crops in one country is known immediately ten thousand miles away; a surfeit of crops

in one country is known with equal prompt-
ness ; prices are regulated not by immediate home demand, but by the supplies and
demands of the world, modified by cost of
transportation and a few other complications well understood everywhere by dealers. The farmer of the present day feels all
this. He has ceased to be able to live from
his farm alone. To maintain his family with
the luxuries and habits which people now
think necessary he must draw upon all parts
of the world, and it is only by selling his
crops that he can pay for what he so purchases. In the older farming districts this
may apply to little more than clothing and
a few unnecessary luxuries, but in the newer
districts it is otherwise. On the prairies
west of the Missouri the farmer raises little
but grain and stock ; everything else, even
the fuel to keep him warm in winter bliz-
zards, must be bought.

The general effects already accomplished
by the manufacture of power may be briefly
stated. Every part of the civilized world

now draws its supplies from every other part of the world. Even the savage or barbarous regions, which get nothing from outside their own country, are forced to render their contributions to civilized lands. Prices are fixed by the whole world, not by any one community, the cost of transportation and tariff charges being simply additions to the cost of the article somewhere else. Furthermore, every civilized family must buy a large portion of what it consumes, and practically all but farmers must buy everything they consume. If the price of wheat rises in Liverpool, it rises on every farm in the United States; if a great surplus is produced in India or Argentina, the price falls in European markets and on every farm in the United States.

It is the glory of the new epoch that it has reduced the cost of almost every article of important use in modern life ; it is the crowning merit of the nineteenth century that it has made cheap so many things which in our fathers' times were dear. The older inhabitants remember the pinched, under-fed look of a large portion of our people, in the early days of the century, when their wages would not provide them with either adequate food or suitable clothing. Such people may still be seen, but there are few of them among the plump, well-clad people of to-day. It is a rarer thing to see children barefoot now than it was to see them with shoes on fifty years ago. Let no one find fault with the lowering of prices of those articles which are necessary for the common support of all, of food, of fuel, or of ordinary clothing.

The manufacture of power has cheapened all the necessities of life ; it has done so by substituting inanimate force for human muscle and strength ; and it has done so by rendering the products of every part of the world available in every other. Great as this work is, it is not the whole ; the change is elevating mankind, and putting the individual in a better position than he was in before. A man's wages are determined, not by what he can do himself, but by what he can have done. The character of his work has risen ; his average pay is better.

One other thing must be observed ; the gain is accruing to civilized nations. Savage and barbarous nations are made to contribute their share towards the comforts of the civilized without receiving much return. This cannot continue. Savage and barbarous people disappear before the stronger arms of the more civilized. A few people are elevated, more of them are crushed. It is a constant struggle for supremacy in which the nation which has the greatest resources, the greatest strength, and understands best how to manufacture and use power, always comes out ahead. To do this it must know what every other land is doing ; it must use the tools which every country has provided ; it must avail itself of the work of all. There was a time when nations could shut themselves up within their own limits and do very well. It amounted to little more than excluding some foreign trade, which at best

was small ; the whole world was nearly sta-
tionary ; the improvements made in foreign
countries amounted to so little that the work
done at home was all that was needed.
Japan lived in this way for centuries, with
laws, practices, and currency entirely unlike
anything now in the world. Gold and silver
were kept in circulation together at a valu-
ation in which gold was only four times as
high as silver. The manufacture of power
has rendered everything like this absolutely
impossible. There is no race or people so
great that it could afford to shut itself out
for ten years from what is going on elsewhere. The simple result would be that
after a decade of stagnation it would find
itself remanded to a secondary position
among the nations of the world, and become
a servant where it should have been a master.

The lessons of history must be studied
as showing the mistakes of the past, not as
giving precedents to be followed now. The
works and doings of the past are not those
of the present. History gives us a record of

what has been done, but no more. It would
be as wise to cite the habits of savage life
as the ways which civilized nations should
follow, as to make the practice of the beginning of this century, before the effect of
the manufacture of power had been felt, the
standard of the present day.

In the race which we are now in the midst
of, which the manufacture of power is opening, new influences, new appliances, new
powers, and new forms of education appear
every day ; only by constant effort, constant
intercourse, continual study, and vigorous
achievements can any country use the tal-
ents which are now before it. If the Americans are to make the best use of their
America, they must call to their aid the
work which the brains of Europe, and before long those of Asia, will contribute to
the general benefit of mankind.

There are other laws than those which
are enacted by legislative bodies, whether
those laws be expressed in congressional
enactments, in judicial decisions, or in that

organic law which has come to be called constitutional. In history the laws of religion have perhaps played a greater part than the civil laws of states. Many of these religious laws have been the work of designing priests or wild impostors, but the preaching of Peter the Hermit carried the young men of Europe to Syria, and the trumpet of the Prophet spread Islamism over the early home of Christianity. There are the laws of logic and mathematics, which are absolute and fixed ; it is beyond any power, either human or supernatural, to set them aside ; not even the Deity can change the sum of two numbers. The laws of trade are at least as important as legislative enactments ; they involve the arguments of logic, the truths of mathematics, and a knowledge of the various natural conditions which have made the trade of the world possible. From their very nature these laws must be international, except in some country which would set itself up alone and sell its nineteenth-century birthright.

III

CAPITAL

THE human mind has been defined as differing from that of animals in being the only mind which can make thought an object of thought. This is the metaphysician's method of looking at it. From the point of view on which our argument has been developed, the difference between man and the lower animals may be said to be that man alone has the capacity to use tools. Fire, the bow and arrow, and pottery were among man's earliest tools. The use of powerful tools is characteristic of the development of the new epoch. Every variety of tool is substituted for the muscles and animate forces which were used before. The whole history of the development of man from his primitive weakness has been the history of the development of his tools. Furthermore, the new epoch differs from all preceding epochs in that it uses tools of great size and cost. Several hundred thousand dollars have been spent for a single steam hammer, a single hydraulic press, or a single train of tools, in a modern steel plant, and each of these is but a single machine among a multitude of others necessary to equip the whole mill. More than two million dollars has been paid for a single passenger ship. The aggregation of many freight steamers makes the fleets which have so greatly reduced the cost of water transportation. The great railroad systems are simply tools for land transportation. It is only by the use of these great tools and by great combinations of smaller tools that the element of human labor has been so largely eliminated from the cost of production and the extraordinary accomplishments of the new epoch made possible. Though the manufacture of power gives practically unlimited power where and when we need it, the cost of the machines to manufacture and use that power must be in some measure pro-

portionate to its amount ; but the larger the
tool or the greater the combination of smaller
tools, the less the amount of human labor required to direct these tools in proportion to
the power which they develop or use. This
is the great secret of modern manufactures ;
it is this which has enabled the fairly paid
labor of northern Europe to fill the markets
of Asia with manufactured goods, although
the workman in Asia may earn only a tenth
as much as the European ; it is this which
now makes it possible for the United States,
paying the highest wages in the world, to
compete with European manufacturers in the
most important products, and has made the
price of steel about one-fifth less in Pittsburg than in Glasgow.

It is not many years since a furnace
which would produce a hundred tons of pig
iron in twenty-four hours was considered
large. The new plant of the Carnegie Steel
Company at Duquesne, near Pittsburg, consists of four furnaces, each of which produces a little more than five hundred tons of

pig iron in twenty-four hours, the whole out-
put being over two thousand tons, with the
result that the bill for labor at the furnace
has been reduced to about six per cent. of
the value of the metal ; but the cost of this
plant, with all its machinery and labor-
saving devices, was measured by millions.
The same may be said of nearly every operation by which the full advantages of the
new epoch are realized. Great results in
economy and production are only obtained
by a thorough equipment to produce them
and by a correspondingly large investment.

Large aggregations of capital have become a necessary feature of the new epoch.
In no other way can the tools and appliances necessary to develop its capacity be
obtained. This aggregation of capital means
one of two things, — either the concentration of great wealth in the hands of individuals, or the collection of the wealth of
many in corporate ownership. The two are
gradually becoming combined ; individual
manufacturers are generally availing them-

selves of legal provisions to place their
affairs under the protection of corporate
ownership; there is nothing to prevent the
concentration of the greater part of the
stock of a corporation in the hands of a
few men or even a single individual. This
tendency, however, results in a concentration of control and management rather
than in a real concentration of ownership.
However great the wealth of individuals
and the amount of property concentrated
in a few hands, the wealth of these individuals may be much less in the aggregate
than the savings and small capital of people
of small or moderate means; and when
these people are working for wages, as
nearly all of them are, they have no use for
their savings or small capital. This small
capital, though the owner may not understand what really becomes of it, is used by
the corporations and by the wealthy men
who are carrying on the great manufacturing, transportation, and other active works
of the country. The capital of the nominal

owners and the nominal capital of corporations is generally very much less than is required to conduct the business. The other money is borrowed ; borrowed from small investors, from savings banks, and other financial institutions ; borrowed at rates of interest which are low in proportion as the security is large. All of this borrowed money has a claim on the capital of the corporation and on the wealth of the individual manufacturer before any profits can be distributed. The apparent owners may take the larger profits, but they assume the risks ; the ultimate ownership may be said to be with the owners of the borrowed money, who must always be protected, even at the entire loss of every other interest.

Theoretically it would seem that the ideal method of concentrating capital would be by coöperation, so that the small capitalists and the operatives might be the actual and responsible owners, entitled to divide among themselves the profits of the enterprise. It is possible that some method

of accomplishing this result will ultimately
be worked out ; if it has been done suc-
cessfully heretofore, it has only been on
such rare occasions that they have little influence on the general conduct of affairs.
But though apparent coöperative ownership may not exist, real coöperation does.
It exists through the agencies of savings
banks and other similar organizations, which
receive as deposits the savings of the workers and lend·them out to swell the capi-
tals of the producing organizations ; it exists
in the systems of preferred securities, as in
the case of railroad and other corporate
bonds, which are entitled to their interest as
an absolute charge, or in the case of preferred stocks, which are entitled to their
dividends before any profits can go to the
common stockholders. The results are very
far from perfect, and have by no means
reached their full development, but the real
ownership is much more widely distributed
than the workers generally know.

The management of these corporations is

one of the principal duties of the new epoch, and the duties of such management are entirely changing the conditions of business success. The work of the manager is to handle tools ; it is to manufacture rather than to trade. While the products must be sold, the greatest skill must be shown in getting the largest results of which the tools are capable. In old times fortunes were generally made by people who had some special advantage over others in the information which they possessed. Such was the case with merchants trading with distant countries. They bought at low prices in the East goods which they sold at high prices at home ; they kept to themselves the difference in these prices ; the distant people from whom they bought knew nothing of what they did with the goods; the people at home to whom they sold them had no idea of what the goods cost. It was the general characteristic of the merchant, from the earliest times, to keep his knowledge secret, and to make the best bargain he could. This method

of conducting trade is virtually at an end ;
the conditions of the new epoch render it
impossible. There will soon be no secrets of
trade by which these profits can be made ;
knowledge of what is going on anywhere in
the world will be open to all, and any one
who desires to do so can learn any specific
thing. The profits of the new epoch must
be made, not by buying cheap and selling
dear, but by reducing the cost of production. The most successful man will not be
the one who has the shrewdest salesman to
dispose of his goods, but the one who can
manufacture his wares more cheaply than
any one else engaged in the same work.
The most successful transportation line will
not be the one whose agents are most active
in securing business, but the one which is
the most closely handled, which can carry its
freight at a less cost to itself than any com-
peting line. Permanent success will depend
not on commercial drummers, but on the
civil engineer ; not on the shrewd guesses
of the so-called business man, but on the

accurate knowledge of the manager who knows what his tools are, who knows what it costs to produce, who knows the defects of his plant and the features in which it may be improved, who in fact is applying all the intelligence of an educated mind, not to getting the better of some other man who may know a little less, but to getting the best work possible for himself and his employers out of what he has to work with.

The management of the corporations which will own the great tools of the next century seems likely to be the great duty of the active man, of the man who is to make the records of the future. As these tools become larger and the demands upon them greater, the responsibility and the honor which belongs to the successful management of such interests will become more fully recognized. The sharp mercantile spirit must gradually go down ; it is already doing so. Careful skilled management is the only thing which can produce permanent results. The business of the new epoch will be manufacturing and banking rather than trading. Good and bad times must be expected, but they will be less common as business trickery decreases and as the world attaches less weight to commercial guesses. But a proper management of corporations must recognize the existence of such periods of prosperity and depression ; it must recognize that while its own work is permanent and not temporary, the demand for its products is in a measure variable ; it must remember that the owners who always insist on complete division of immediate profits must ultimately fall behind, and that the extra profits of the good times must be used to maintain the plant and put it in a condition to tide over bad times. Every manufacturing plant, every steamship, and every railroad is simply a tool. It has no value except in its capacity to produce ; improvements are constantly being made, and as these improvements are made, the parts which are superseded must be discarded.

The relations between the ownership and

the management of corporations have not
yet been settled. The new epoch opens as an
era of corporations, and these relations are
gradually being fixed. The old rule of equity,
that any one handling the property of another
was a trustee and had no right to use that
property for his own benefit, is fundamental, and must always govern the management
of corporations. In a corporation there are
three elements at work, — ownership, management, employees, — the ownership including not only the nominal owners but all
the ultimate ownership, which often extends
to the employees. Ownership is very much
more scattered than commonly appears.
The employees are the mechanics and la-
borers who handle the machines and do the
comparatively small amount of manual labor
which machinery has not yet mastered. The
manager is the trustee for both parties ; he
must understand the tools that he uses and
get the best work out of them that he can
for the owners ; he must also understand the
relations between those tools and the men

that work them, and see that the workmen are competent and cared for. The rights of ownership are the rights of property, which seem to be the foundation of all civil government. The manager is the trustee for this ownership. The rights of the employees are the rights of men and women who are profiting, and must be allowed to profit, by the developments which the manufacture of power has rendered possible, and to whom must ultimately accrue the increased leisure and comfort which the substitution of mechanical power for muscles renders possible. The owners and the workers both have rights ; the managers have principally responsibilities.

There is another feature which must not be overlooked. The necessary aggregations of capital and the necessary extensions of corporate ownership and management will have an important effect which many may deplore, but which seems to be inevitable. It is one of the changes which necessarily occur in the passage of races from one stage of civilization to another, in which some of the best features of the older life pass away, leaving a deplorable void, while the new conditions, with their new virtues, are being developed. The new epoch must reduce the number of people who work by themselves ; it must reduce the number of people who seem to be entirely their own masters, and who pursue their lives on independent lines, and it must increase the number of salaried employees, the number of men who are working for fixed wages, and who are apparently dependent on others. Its development will wipe out the small manufacturers ; they cannot compete against the great concerns. It must wipe out many of the small traders and dealers whose business has always been that of middle-men. In fact, these changes make men members of communities, members of working organizations, members of what has been called an army of workers, but necessarily deprive them of many of the conditions which have hitherto done most to produce strong individual characters.

Were this all it would certainly be an evil. The loss of independence is a great loss; it is always regarded with regret. Whatever his vices and wickedness, and they have been enormous, there is a noble element about the life of the savage man which will live in literature and song long after his sins are forgotten. The strong features of New England country life, with its combination of manual labor and high intelligence, which bred the men who started the Revolution, call for the highest admiration and praise, in spite of the physical pain and suffering which wore out so many of these people before they had passed middle life. But no one would voluntarily go back to the conditions under which our grandparents were born; still less would he adopt the life of the noble savage. It is premature to say where the compensation for the loss of individualism and personal independence will be found, but we may be sure that such compensation will come. It may be in the greater amount of leisure which all men will have when human labor is reduced as that of machines increases. It may be that it will be found in the higher degree of skill and education which must characterize in these days of machinery all classes of labor. It may be that it will come from the general benefit which attends the immolation of the individual in the whole and enables him to recognize that he has a share in what every one else is doing. Whatever form this compensation may take we may be sure that it will come. It may be only after a severe struggle, but the laws which control the development of the new civilization are inevitable and cannot be resisted. The future good of our race lies in utilizing them to the utmost possible extent, and not in trying to retain the good features of conditions which are passing away.

IV

In the new epoch there must be a great increase in the variety of the duties of governments. There will be a great extension of the geographical limits over which a single government can exercise immediate and direct control.

In early times the functions of government were really but two : the protection of the people against foreign enemies, which meant the conduct of war, and the protection of the people from domestic enemies, which meant police and the whole system of both criminal and civil law. This is not the place to amplify these duties nor to consider the different forms of government which have performed these duties. It may, however, be noted that the foreign and domestic functions, even of despotic governments, have usually been so distinctly marked, that when a government has failed entirely in protecting itself and its people and passed out of existence before a victorious enemy, the domestic provisions have been recognized by the conqueror and carried on with little change. The right to carry on war and the duty of protecting and governing the people carry with them the right to provide the means of doing so. This right includes in war the right of the victor to levy an indemnity on the conquered, a right which was formerly held to include absolute confiscation, reduction to slavery, and even extermination. This right in domestic affairs covers taxation in its various forms and with its various substitutes; in its broad application it includes the right to collect revenue and the right of conscription. The right of conscription in modern civilized nations is confined to military service, but it is a power which has been extended to forced labor on all varieties of public works. Incidental to this, though it marks a more advanced condition, is the right to coin and issue money.

These were practically all the earlier functions of government, though other rights and powers have often been assumed, especially by the more despotic forms. Furthermore, governments have usually been more or less linked with religions, a relation which has been one of the most potent influences in history the influence of the priests on the rulers having often exceeded the power of the rulers over the priests, and the duties of education and religion being often confounded.

A departure from old limits of government and recognition of a really new function is the post office, whose introduction preceded the dawn of the new epoch by two or three centuries, but whose general extension to communication with all parts of the world for the benefit of every member of the community is of a very recent date. It seems hard to recognize that there were no mails in the Roman Empire, and that even

to this day there is no general government
mail service in the Empire of China. Two
thousand years ago the young wife who
followed her husband to a distant colony,
whether one of the Greek colonies on the
Mediterranean or at a later date the out-
posts which Rome scattered much farther,
bade farewell for life to all her friends and
relations, from whom she might never hear
again. The change of old friends and surroundings for new ones was absolute and
complete. To-day a European emigrant who
settles in America, in Australia, or in south-
ern Africa, is in constant communication
with the home from which he came; and this
constant communication, at once keeping
up the friendships of the old and gradually
bringing them into contact with the friendships of the new dwelling-place, must have
great influence in unifying and consoli-
dating relations which are of much more
than a personal character.

Another duty of governments, not wholly
new but yet distinctive of the new epoch, is

that of sanitation. Something of this kind has been done in cities for an indefinite period. It was especially marked in the days of Rome, where the sewer built in the time of the Tarquins still carries the drainage of the Eternal City to the Tiber, and where aqueducts built in the days of the emperors still supply far better water than is often to be had in Europe. But with the increased knowledge and capacity of the new epoch, the duties of the government in caring for the health of the people have become very greatly increased. Water supply is not confined to the more important cities, but is extending even to small towns. Water supply must always be followed by a sewerage system, and the modern sewerage system must not only take away the sewage, but dispose of it and utilize it. Other duties follow. They are being assumed perhaps too rapidly. Lighting public ways is a duty of cities, and it may be that the supply of light and heat for private consumers may yet be considered a government duty. In fact, the extensions of the duties of government in the new epoch are too general and too great to be definitely predicted.

Another duty which governments are assuming is that of education. In old times this duty was only recognized in the form of a state religion ; and though the priests were the custodians of literature, science, and art, as they were then understood, it was not their policy to disseminate education widely. The importance of general education, especially in those countries which had free governments, was appreciated before the dawn of the new epoch. It may have had much to do with the invention which led to the manufacture of power ; but be that as it may, the substitution of mechanical for human power calls for general education. It leaves more time for the average man to study, and it demands a higher order of intelligence in the man. The provision of an education has been assumed to be a proper function of government. It began with the free public-school system, which is

now universal in our country and is extending around the world. Its next development
was the library; and it will not be long before every town considers it as much its
duty to provide a free public library as to
provide free schools. A further development
is in progress, and boards of public education are beginning to provide free lectures
to supplement the school system.

The promotion of trade has been assumed
by governments, and this involves the care
of harbors, improvements of navigation,
and many other engineering works, which,
though perhaps originally justified on military grounds, have gradually been extended
until their development for peaceful trade
is much greater than for ships of war.

The tools of the new epoch are already
being used in the conduct of foreign affairs,
and the relations of one nation to another
are being handled at home, the foreign repre-
sentatives being little more than the agents
through whom telegraphic and other dis-
patches are sent. The possibility of geo-

graphical extensions of governments in the new epoch is even greater than the addition of new functions. It is absolutely essential to a satisfactory government that no long interval should elapse between the need of action and action itself ; between the issuance of instructions and the execution of such instructions. Various methods have been adopted to overcome this difficulty. The delegation of power to viceroys and others was the method in use in the empires of old and in Asia to-day; the method of representation came later, and was the only method by which a single popular government could be extended over any considerable area of country. The Constitution of the United States preceded the opening of the new epoch, and its fundamental feature is that it leaves the execution of all local government, which demands quick action, in the several States, which thus far are considered independent nations, while it provides for a representation of every community at the centre of the general government. It is the most perfect device ever made for the extension of a government by the people over a great area of country. The manufacture of power, however, has virtually annihilated distance. With the present condition of steam navigation it may be said that there is no seaport in the world which cannot be reached from any other seaport in a period not much exceeding thirty days. It is true this could not be done by any regular line of transportation, but it could be done in an emergency by the use of special ships. The railroad system has now brought every place in the United States, excepting Alaska, within less than one week of every other place. The combination of the steamship and railroad service may within another century make it possible to go from every place to any other place on the planet in about three weeks. The telegraph goes beyond this, and it is now not only possible but the actual practice to know in every civilized city on the earth what has taken place on the preceding day in every other city. These changes have removed the conditions which formerly limited governments. If a single government could be organized to handle the affairs of the entire earth and give equal rights and like laws to all races and conditions of men, the physical difficulties in enforcing the acts of such government would be much less than those of handling the thirteen States at the time of the adoption of our Constitution.

Whether such universal government will ever exist is a question with which we have nothing to do. So far as physical conditions are concerned, the manufacture of power makes it possible that the world in the new epoch may be governed in this way; whether it ever will be, and whether it would be desirable for it to be, are different things. It seems more probable that the increased respect which different nations will have for each other as they come to know each other better, as each one learns to adopt the better features of others

and to discard its own worst features, may
lead to the existence of a few great nations
which will manage their own affairs separately, but which will bear to each other relations like those which exist in a peaceful family. At present the obstacles to the
unification of the human race are great;
they are an inheritance from the past which
the new epoch will gradually obliterate.
The first of these is that of race. Civilized
races differ from each other. The civilization
of Asia is the oldest in the world, but the
two most populous districts of Asia, China
and India, are separated by a mountain bar-
rier more impassable than any ocean, and
their civilizations were on independent lines
and entirely unlike. The civilization of Europe came later, but it developed with no
knowledge of what had been done in Asia.
Each of these civilizations represents a
development from savage life, but they
are on different lines. The races have civilized apart, and each of them differs from
the others more than either does from the

absolute savage. All the differences, great
as they are, may disappear when the races
come to know each other thoroughly. Another obstacle in the way of unification is
language. This also will in time disappear.
A third great difference is religion, the
power of which has been enormous in all
ages of the world; and while fetichism and
all merely idolatrous religions seem to disappear at once before their spiritual con-
querors, the great religions which have real
life and spirit in themselves, and which ap-
peal to the better qualities of the mind and
soul, do not yield to each other, and the
lines of distinction are to this day as plainly
marked as ever. Whether this division is
to continue with the knowledge which the
new epoch is opening, remains to be seen.
If it does continue it must be accompanied
by that tolerance of each other which should
ever mark the relationship between intelligent educated people.

These mental and intellectual obstacles
will be much more difficult to surmount

than the purely physical obstacles, but if they ever are surmounted the conquest will be complete, while the mountains and oceans will continue, lines of past demarcation overcome by the work of the engineer, but still remaining to help men to understand the history of the ages which are just entering their final period.

It is unwise to predict too much, but we may be sure that the effect of the new epoch will be to reduce greatly the number of nations in the world; that these nations will ultimately be large but compact, perhaps not more than a dozen in all; that another effect will be to increase the duties and responsibilities of government; that there will be great changes in the forms of government, though what these will be we cannot tell; that the needs of the people who are to live in the new epoch, wherever they may be and whatever their form of government, will be discipline and education, — the discipline which, whether in the individual, the family, or the nation, preserves that control under which alone the best results can be obtained, and the education which alone fits man to direct the great powers of the new epoch for the best good of his race.

V

CIVIL ENGINEERING

THE new epoch has opened an entirely new set of professions. The old professions were primarily divided into two classes, the military and the civil, and of the latter only three were recognized, — divinity, law, and medicine. These three were called liberal professions, and their members were supposed to be, and generally were, better educated, though not always more thoroughly trained, than the men who followed other callings. The demands of the new epoch are such that educated men are required everywhere. They are needed to design the tools by which power is manufactured and is utilized ; they are needed to manage the affairs of the corporations whose capital is invested in the great variety of tools, and which have been referred to ; they are needed to perform the increased duties which governments are now assuming.

Seventy years ago civil engineering was defined as the art of directing the great sources of power in nature for the use and convenience of man. [1] This definition was embodied in the charter of the institution which has done more than any other to unite the profession and to give it the standing it is now attaining. It was made in the very infancy of the new epoch, within sixty years of the time when Watt developed his first steam engine. Had the profession remained unnamed till the end of the century, it is possible that its various departments might have been classed separately, and that what is now called by a single name would have been divided into several professions. The definition was followed by a list of objects and applications, but it was expressly stated that its real extent was limited only by the progress of science, and that its

[1] Thomas Tredgold, 1828; subsequently embodied in charter of the Institution of Civil Engineers.

scope and utility would be increased with every discovery, and its resources with every invention, since its bounds were unlimited, as must also be the researches of its professors. This definition is broad enough to embrace every department of work which undertakes the development and use of any of those physical powers through which the new epoch is now subjecting all varieties of matter to the dominion of mind.

The constitution of the American Society of Civil Engineers fixes as a requirement for full membership "the ability to design as well as direct engineering works." The English definition and the American requirement taken together explain what constitutes a civil engineer. His business is to design the works by which the great sources of power in nature are directed. His works are not built for themselves nor as commemorative monuments ; they are made to direct the powers of nature for the use of man. Every engineering work is built for a special ulterior end ; it is a tool to accomplish some specific purpose. Engine is but another name for tool. The business of an engineer relates to tools. A civil engineer must be capable of designing as well as handling tools. The highest development of a tool is an engine which manufactures power. All the great possibilities of this profession come from the existence of such tools.

It is a common error to think of civil engineering as a coördinate branch of a general profession with many other branches ; to class it with mechanical engineering, with hydraulic engineering, with sanitary engineering, with mining engineering, with electrical engineering, or with any other specific branch. The name of every special branch of engineering has a distinctive meaning ; the mechanical engineer deals with machines ; the hydraulic engineer, with water ; the mining engineer, with mines ; the sanitary engineer, with drainage ; and the railroad engineer, with railroads. The word " civil " has no such distinctive meaning ; it shows only that civil engineering is the work of the citizen and not the work of the soldier. Civil engineering, in its true meaning, embraces every special branch of engineering. The professional limitation which should be applied to the civil engineer is that he must be a man who in his own department can design as well as direct. He must have that control over his work which nothing but intelligent knowledge of the subject gives. He may be a railroad builder, he may be a skillful surveyor, he may be a mechanical engineer, or he may follow any other specialty; but whatever he does he must do it not as a skillful workman but as one qualified to design. Any man who is thoroughly capable of understanding and handling a machine may be called a mechanical engineer, but only he who knows the principles behind that machine so thoroughly that he would be able to design it or to adapt it to a new purpose, whatever that purpose may be, can be classed as a civil engineer. Any skillful sewer-builder and pipe-fitter may claim to be a sanitary engineer, but only the man who approaches his work with the intelligent knowledge of the conditions which sanitation involves can be classed as a civil engineer. Any man who knows how to work a mine may be a mining engineer, but only he who understands why he works his mine as he does can be called a civil engineer. Any well-practiced electrician may be classed as an electrical engineer, but only one whose practical knowledge is based on the intelligent study of electricity can be called a civil engineer. The business of every engineer is to handle tools ; the business of the civil engineer, whatever department or specialty he may follow, is to design and build those tools rather than to use them. The word " tool " is used in its largest sense ; it may be called engine if preferred. Any constructed thing whose principal object is to produce something outside of itself is a tool, whether that tool be a brad-awl, a steamship, or a railroad. Civil engineering embraces all branches of engineering, but the civil engineer differs from other engineers in that he makes tools rather than uses them. The relation of civil engineering to all other branches is of the broadest kind ; no branch of engineering is excluded ; the only exclusion is based on the individual qualifications of the men.

The civil engineer is briefly a man who, with knowledge of the forces and materials around him, uses that knowledge in the design and construction of engineering works. His business is to design the tools by which the sources of power in nature are directed for the use of man. A body of civil engineers should include the choicest minds in every branch of the engineering profession.

Any civil, military, naval, mining, mechanical, electrical, or other professional engineer, architect, or marine architect, who, with knowledge of the great sources of power in nature, uses that knowledge in the design and direction of engineering works, is qualified to be considered a civil engineer. Civil engineering includes all branches of engineering, but not all engineers. It should include every engineer who applies a knowledge of the powers of nature to the design and construction of engineering works. It should include the architect who uses this knowledge in his designs and constructions; but the architect who treats his profession as a fine art to decorate a construction which he cannot design, belongs elsewhere. Intelligent knowledge of the great powers in nature is the fundamental requirement for a civil engineer. On this substructure a superstructure of actual design and construction must be built to make the complete professional man.

The civil engineer of the new epoch, the epoch which he is bringing into existence by the manufacture of power, must be an educated man. In no profession will this be more necessary. The physical laws of power and strength are mathematically exact and admit of no trifling. As the epoch progresses the requirements for each individual will become more complicated. The theologian and the metaphysician may claim that an education based on the laws of matter leaves out the highest part of existence; the biologist and the physician may claim that matter endowed with life is a higher organism than the inanimate matter with which the engineer has to deal. But however true these claims, their laws have not the mathematical rigidity, the clear definition, and the thorough discipline which mark the laws with which our profession works. The engineer cannot shield himself under doctrines or theories which he accepts but cannot understand. Dealing with accurate, definite laws and guided by the corrective touch of physical nature, the education of the engineer will become more necessary, more thorough, and more exact than that of any other professional man. This is the training which the civil engineer of the new epoch must have. This knowledge he must have, or he must be classed as a workman rather than a professional man.

The civil engineer of the new epoch

must sink the individual in the profession. The engineering work of the future must be better work than has ever yet been done. The best work is never done by separate men. It is only accomplished when professional knowledge so permeates all members of a profession that the work of one is virtually the work of all. The first steps are made by individuals, but the best results come later. In the Middle Ages Gothic cathedrals were built throughout northern Europe. They are exquisite works; no modern architect can approach their beauty. The reason is that the men who built the Gothic cathedrals worked together as members of a guild which was thoroughly imbued with the spirit of building these churches. In no period of the world's history has marine construction had any significance compared with what it has to-day, and it is because the great shipbuilders are working together, each having the practical benefit of what they all are doing. They are working together as members of a profession rather than as individuals, and their work is becoming more uniform and more perfect.

The civil engineer of the new epoch must be a specialist. No man can learn to design all the tools by which the powers in nature are to be directed. The work is too great for one man to master. The best results will only be obtained by concentrating effort in a single line. But though the civil engineer must be a specialist, his specialty must not be of a narrow kind ; he must have that general knowledge and training which makes the liberally educated man. In every occupation a natural selection of men takes place ; some follow the close lines of the work for which they are trained, while to others this training is but an incident in the early part of their careers, and does little more than point the general direction of their lives. The ability to deal with men and to direct the minds of men is a matter of natural gift more than of education. It is so important that when possessed in a high degree, all other accomplishments yield to it ; and its possessor, realizing that the ability to use several minds gives him the same advantage among his fellows that the control of additional power has given among races, will use his capacity. But the positive training of an education has its value for men whose paths of life may lie far from the special work for which they were trained ; it will at least teach them the importance of accurate knowledge. Too many men are contented to guess rather than to know, relying on personal tact to relieve themselves from difficulties when their guesses are wrong.

The civil engineer of the new epoch must fill many positions which are now held by men of different training. The knowledge of the tools, both large and small, which men are using, must be the strongest qualifications for their use. Accurate engineering knowledge must succeed commercial guesses. Corporations, both public and private, must be handled as if they were machines, and the men who will so handle them will find their best training in the education which will make the best civil engineers. These managers may not be called civil engineers, but civil engineering should not find fault with titles; the man whose training has fitted him to do the work of a civil engineer will not cease to be one if he is promoted to a high position of management.

When ability to rule meant ability to defend against invasion, to maintain war against foreign enemies, government was in the hands of soldiers. As society became more complicated, and a permanent administration of civil matters more necessary, domestic affairs being more important than foreign, the administration passed largely into the hands of lawyers. The legal profession was long the only educated profession whose members were available for public work. The functions of government are changing. The demands of the new epoch are not like those of the past. Safety from

foreign invasion is needed less than safety
from dangers which lurk within, — from the
poisons, both moral and physical, which en-
danger concentrated population ; from bad
air, bad water, and bad construction ; from
corrupt administration, and from bacteria.
The rulers and governors, who at first were
soldiers, who subsequently were selected
from men trained as lawyers, must in the
future be taken, at least in part, from those
who are educated in the utilization of the
powers in nature, — from civil engineers and
the men who are equipped with the new
education for the benefit of their country.
The duties of municipal government, or the
government which is most closely concerned
with local affairs, must become very much
like the management of corporations. In
fact, a municipality is a public corporation
rather than a government, and its duties
should be performed in the same way as
those of a corporation. The same class of
men will do the best work for a city that
will make the best managers of manufac-

turing corporations. In cities and in many communities the duties of the government rest more on good engineering than on legal skill. The whole life of the community depends on appliances and conveniences which the manufacture of power alone has made possible. For all this work the government needs neither soldiers nor lawyers, but men educated in the various departments which come within the broad definition of the work of the civil engineer.

The tools which civil engineers have to build are generally large. The physical man is often a tiny thing beside the work which he has to construct. Nothing better illustrates the power of mind over matter than the work of this profession. Though it deals with matter and its work is of a material kind, it is the mind which has made this matter give forth power ; it is the mind which is opening the new epoch, and it is by the training of this mind that the civil engineer must prevail. He is the priest of material development, of the work which enables other men to enjoy the fruits of the great sources of power in nature, and of the power of mind over matter. He is the priest of the new epoch, a priest without superstitions. But if this profession is to do the good work of which it is capable, the true spirit of individual immolation which has characterized the devoted priest of all ages must be found among its members. The profession can only do its future work by trained minds working together.

VI

THE UNIVERSITY

GREAT as is the effect which the development of the new epoch are having on the engineering profession, their influence on education is equally important. The duties of universities are being entirely changed. Great changes impose new duties on the institutions which are charged with the intellectual development of the community. No changes have ever equaled those through which the world is passing now. No institution has greater responsibilities at this time of change than those which rest on a university. The manufacture of power has an intellectual as well as a physical effect; it has separated power from the mind which must manage it; it calls for intelligent design and direction of the multitude of works which it has rendered possible; it has equipped our generation with tools for study and investigation as well as for mechanical work.

A university is more than a school; it is not merely a college; still less should it be an eleemosynary institution for the benefit of young men to whom it can give an education. A university owes its duty to the community as a whole, not to individuals who live in that community. The endowment which a university may receive, whether it come from public appropriation or from private gift, must come to it as to a public benefactor, endowed and sustained in order that the whole community may have the benefit of its intellectual guidance. It must not train young men because those young men wish to be scholars, but because trained scholars are necessary for the good of the community. The individual must be sunk in the nation or state of which he is a part; the young men whom the university educates should know that they are educated to be useful members of a community, and not for their own ends. The real duties of a university are universal; it is the head of the educational system of the land, charged with the high responsibilities which this position implies; it must be the depository of the lore which former generations have accumulated and the pilot of the community in every kind of intellectual life; it must preserve the records of the past, and it must train the men who are to make the records of the future; it must combine the work of a museum with that of a school.

A collection of physical objects, though those objects be most rare and curious, does not make a museum. A collection classified and arranged in the most systematic manner that has ever been devised would still be incomplete. It must be a collection of the records of the past, including that which can be stored only in the mind. A classified museum, though it include a library containing every book that has ever been written, would be of no value without the minds to use it. The museum which forms so important a part of a university must include among its collections a collection of educated men.

The school which is to train the men who are to make the records of the future must build its special courses on the foundation of an education which teaches how to use the mind. This is the real measure of a liberal education; without this, the men it educates will be of little value in the community.

The new epoch which the manufacture of power is bringing forth makes new demands upon a university — new demands upon it as a museum in the large sense which has been stated; new demands upon it as a school to train the young men whom the community needs, and who will make the records of the new epoch.

The new epoch has an inheritance from older times. It increases the work of a university in its capacity of museum. In the mere collection and preservation of records, the work is greater in a period of change than at any other time. Generally, in passing from one ethnical period to another, the records of the past have been lost. The students of the earliest life of man have to grope among prehistoric remains, deciphering marks which seem almost as inanimate as geological strata, and tracing their uncertain way by analogies drawn from races living to-day.

The new epoch must destroy as well as build; the new civilization will wipe out the conditions which precede it. The savage and barbarous tribes which now live simultaneously in different parts of the world must disappear. If their habits, customs, and mental conditions are to be recorded, the work must be done soon; in one or two centuries it will be too late. The structures which represent the achievements of many generations cannot be preserved. A few may be kept as beautiful relics, specimens in a universal museum. But the manufacture of power has made

the demands of the new epoch so different from those of the old that nearly
everything which has to be used must be
built anew. The old and the new cannot
exist together. It is hard to realize how
rapidly the appearance of the whole earth
may change. Greater care of life is a
feature of the new epoch. An increase of
population at the rate of one per cent. an-
nually, which is less than that in European
Russia, would cover the entire land surface
of the globe, including deserts, mountains,
and snow-capped plains, with a population
as dense as that of Belgium, in about three
centuries. In the change through which
we are now passing, a change which will
leave no isolated tribes for the future, it is
one of the duties of the university to see
that the museums of the future are stored
with the full history of the past.

The new epoch is characterized by great
material changes. In such a time there is
danger that natural science and physical
study will overpower all other thought.

The treasures of philosophy, of music in the broad Greek meaning, and of religion in the noblest sense, must be a special charge of the university.

Around the museum, of which they will form a part, must be gathered the men who will collect, study and care for what it contains. The university must train and educate these men to be the curators and scholars who will see that record precedes destruction ; who will take care that, when physical existence ends, the facts which scholars need are preserved, — and who will themselves be the scholars who are to use these records. The education of these men must include the intelligent study of the delicate accomplishments and refinements of the past ; the new epoch may not have the grace and taste which have marked some inferior conditions ; in the creation of beauty, Europe and America are to-day far below the nations which dwelt around the Mediterranean two thousand years ago, or the older races which still inhabit Asia. The study of history belongs to this department. The training for those professions which are based on history and precedent will find a place here. But few of the young men so educated will remain to form the body of educated men which is an essential part of the university museum; the majority will seek other lives and callings. The general body of educated men, as education was once understood, the men who are students rather than workers, readers rather than originators, who are guided by what others have done rather than by what they themselves would do, will be educated in this museum.

This work is much like what the university has always been doing. In this department the effect of the new epoch is to develop rather than to change; it makes the old work greater and more important than before; greater because there will be more workers, more to do, and more tools to work with; more important because much which cannot be done soon may be lost for-

ever, and because the life of a community
busied with material development needs a
double leaven from the educated past.

But the community has needs for the future as well as for the past. The records of
the past must be preserved and studied by
that body of educated men who make the
society of a university town the most refined
and intelligent that is anywhere found, and
who give to the precincts of a university a
peculiar attraction which exists in no other
place. The records of the future must be
made by men of different types and different
habits, who are educated to fit them for
active work, who will exchange the pleasures and quiet of the university for the roar
of the rolling-mill, the buzz of the machine-
shop, the obscurity of the mine, the bustle
of the railroad, and the harsh surroundings
of many other duties. These men must be
prepared to sacrifice the pleasures of education as such, and the delights of study for
mental development, and spend their lives
where their work calls them.

The definition of civil engineering which
is incorporated in the charter of the Institution of Civil Engineers has already been
quoted, "the art of directing the great
sources of power in Nature for the use and
convenience of man." The same definition
may be accepted as measuring the duties of
the new education which is to train young
men for active work in the new epoch ; this
education must qualify them to handle all
the great sources of power in nature,
whether those sources be animate or inanimate, whether the direction be mechanical or physiological, whether the work be
investigation, construction, management, or
invention ; it must be prepared to deal with
every kind of matter of which the world is
composed, with the power associated with
such matter, and with the laws, simple and
complicated, which govern it ; the object
must be to direct such matter and power for
the improvement of mankind ; this must be
the work of the new education. The civil
engineer claims that all this work belongs to

his profession, which should include every
educated man who, with a clear knowledge
of the laws which govern his work, is handling the powers of nature, be that work in
a harbor, a machine-shop, a railroad, a mine,
an edifice, or a laboratory ; the fundamental
condition being that the work shall be that
of an educated man, who knows how to
design and to direct, in accordance with
nature's laws of construction, strength, and
power.

There is one profession whose age and
history have given it a rank by itself. Medicine had an old and honored name when
civil engineering was still unrecognized. But
it belongs with the new profession rather
than with the older ones ; its work deals with
the powers in nature for the use of man.
It differs from engineering in that it deals
with organic life, and not with inanimate
power. Its recent developments have been
rendered possible by the same conditions
which have developed engineering. Its place
in a university is with the other branches of

physical science in the new education, rather than in the historical museum.

The time is not far behind us when none of the occupations which strove to direct and use the sources of power in inanimate nature required any high degree of education. Practice and experience seemed to be enough. Good sense, guided by precedents, accomplished what was necessary. While in some ways a man specially educated had an advantage, it was not enough to give his work the marked position which belongs to an educated profession. This is no longer so. Within the last half century the whole conditions have changed.

It is not the educated character of the man, but the educational needs of the work which makes an educated profession. The work must be such that it can only be done by those whose education has specially qualified them for it. Natural ability combined with education will always be greater than either of the two alone ; but no occupation can become an educated profes-

sion until education gives the men who follow it a distinct advantage over those who
have not received such education ; and no
profession will ever be composed entirely
of educated men until the advantages of
education outweigh those of mere natural
ability.

The manufacture and use of power, though
in its crude beginning easily understood and
handled, has already reached a point where
accurate knowledge and thorough training
are needed for the best results. There is
not a single department in the manufacture
or use of power in which the advantage of
a thorough education is not felt.

The study of the strength of materials,
and the mathematical laws involved, is required in all structural work. The older
structures were the gradual development of
experience, each builder inheriting the work
of his predecessors. So long as dimensions
were small and the material generally exces-
sive, this worked well, but modern engineering asks for the least material which can be

used to produce safe results ; the strains in every part of the structure must be calculated, and unnecessary material removed ; the rule that nothing is stronger than its weakest part must be applied by eliminating the material which gives useless strength.

Metallurgy has become in all its details a matter of refined investigation. A minute variation in the amount of phosphorus it contains will make the difference between a bar of steel which is perfectly safe for structural purposes and one which is treacherous and may break without warning. A large portion of the steel product of the world is now made in furnaces with basic linings which absorb the excess of phosphorus, and which were introduced, not by a practical iron-master, but by a chemist, who made dephosphorization his special study, and sacrificed his life to the ardor of his researches.[1]

The ordinary high-pressure slide-valve steam engine, such as is used for a sawmill

[1] Sidney Gilchrist Thomas.

in the woods, or for a straw-burning harvest outfit on a Dakota prairie, is a simple thing which anybody can understand, but its use is only justified because temporary convenience is more important than economy. The marine engine, where power is limited by capacity to carry fuel, is very different; scientific study and design have reduced the coal consumption of the best marine engines to less than a pound and a half per indicated horse-power; this has rendered possible the speed of the modern Atlantic liner and the extremely cheap carriage of the tramp freight steamer.

Electrical engineering, and the other professional branches which are multiplying rapidly, require a like scientific training.

This education is not a simple one. A smattering of knowledge may enable a man to understand what is going on, but to design and perfect the structures and machines which will give the best results requires a thorough knowledge of laws whose complications increase as their applications are extended. The strength of materials, the chemical composition of substances, the laws of heat and of dynamic energy, with other equally important principles, enter into almost every operation of modern life. Every design must be worked out in accordance with the laws which govern it. There was a time when Yankee ingenuity was thought to be equal to anything, and the memory of that time still exerts its baneful influence ; works which required educated engineers have been intrusted to ignorant men, and terrible disasters have followed this perversion of trust. The laws which govern the problems of mechanical and material devices are complete, and require trained minds for their solution ; they are exact ; they can be demonstrated absolutely, and a mistake may be followed at once by a disaster. There is no place among them for the strange theories which, when without the corrective influence of physical facts, seem to prove intellectual depravity ; the man engaged either in the manufacture of power or the utilization of its sources in nature, can find no refuge behind unproved theories or questionable practices.

This work is the creation of an epoch differing from the past to such a degree that it may itself be considered new ; the education which will fit men to perform this work must also differ from the old education. The old education teaches facts ; it is based on a knowledge of what has been done. The new education cares little what has been done, provided no one ever wants to do it again. The men who are to adapt the great powers of nature to the use of man, who are to make the records of the future, must understand the laws by which they are to do this, must know how to investigate, and how to work themselves, rather than know what work other people have done. No work is good unless made on correct principles, and education must equip the worker with these principles. The education of the engineer is intended to fit him to construct and use tools which serve some specific purpose ; they must be adapted to their purpose and nothing else ; he must be prepared to see them thrown away when their work is done. The machine must be properly proportioned ; the heavy, clumsy tool which requires unnecessary power must be avoided as much as the weak tool which fails under its work. Furthermore, this education must be applied to every class of work ; to all that great variety of tools by which the engineer utilizes the powers of nature ; to those more permanent constructions by which the architect would build monuments for future ages.

As this education becomes more general, it will be realized that the basis of all true beauty is adaptation to its purpose ; that the decorations of the so-called fine arts must be made subservient to correct and simple lines of construction, which they would emphasize rather than conceal. The false motto *Ars celare artem,* which really means it is good to lie, must give place to the glorious *Veritas.* The incongruous absurd-

ities of the present day must disappear.
The engine frames of the first Cunard
steamers were decorated with Gothic arches ;
beside the engines of a modern steamer
these old machines have a strange fantastic
look. Architecture, which as a fine art
would consign itself to the museum, and
sometimes, following the rapid changes of
fashion, seems to differ from millinery
chiefly in the want of a beautiful object on
which to place its novelties, will find its
highest development in correct construction.

The engineering of the new epoch must
be thoroughly good. This means the development of the true professional idea, and
demands professional education. The best
work has never been done by separate men ;
it is only accomplished when professional
knowledge so permeates the whole body of
workers that each member has the benefit of
what all are doing. The first steps in invention and in new developments are taken by
individuals; the best work is done later when

the path into which the bold inventor ven-
tured alone is trodden by the crowd who
find it their usual course. The name of
Watt was immortalized by his successful
introduction of the steam engine, but there
are thousands of men to-day who can build
better engines than James Watt could. Marine construction owes its present high condition to the fact that ship-building has
become a profession in which each builder
has the real benefit of what all are doing.
There lived in one of our great cities an
engineer of marvelous inventive skill and
world-wide reputation, who in a variety of
ways has left his mark on the developments
of the century ; his history was a mixture
of great accomplishments and strange dis-
appointments ; but the saddest part of the
whole was the work of the last years of his
long life, when, alone, having little intercourse with other men, he set himself the
task of devising means by which future generations might manufacture their power
when the supplies of fuel now in use should

be exhausted.[1] Perhaps no engineer who
has ever lived was as well qualified to solve
this problem as he was ; but no man, however great, can do good work alone and
before its time. When the problem on
which he toiled for years becomes a real
issue, there will be many men, of far less
ability than he, who, sharing the professional experience which will come meanwhile, will have little difficulty in providing
the needed power.

But the best professional spirit demands
more than this. To training and instruction
must be added the spirit which alone makes
men worthy of the power education gives
them. They must not only know how to
work, but they must do it in the spirit which
the best good of the community demands.
The advance of mankind through the savage
and barbarous periods was not continuous.
Increased powers are susceptible of abuse
as well as use, and the evil of the abuse has
sometimes exceeded the good of the use.

[1] John Ericsson.

The new epoch will be no exception; its universality has only substituted other dangers for the barbarian invasions which destroyed older civilizations. The men who would sacrifice their friends and their country for their own selfish selves still live; the greater their capacities the greater the danger. Never before have the opportunities for selfishness been so great, whether that selfishness be devoted to acquisition of useless wealth, to indulgence in degrading luxuries, or to the general disregard of the rights of others, which may characterize poor and rich alike. In communities where everything is organized on the selfish basis of commercial life, these influences may transform the great forces of the new epoch into powers of destruction from which the world will never recover.

There is a capacity in the mind which can be developed to meet these dangers. The antidote for these evils which selfishness begets is that power which, working in many ways and for many objects, takes a man out

of himself and is called love, whether that
love be for human beings, for animal life, for
inanimate objects, or for laws and principles,
which are at least as real as anything else.
The education of the men who are to do the
work of the new epoch must not only train
them and teach them, but must fill them
with that interest and enthusiasm which en-
genders love. This can be done ; the more
complicated the work and the higher the
education, the more interest the worker
finds to make him love his work. Every man
who has entered earnestly into the study
of the powers of nature, into the design of
works which are to utilize those powers, or
the execution of the plans which the world
is profiting by, knows that this is true. The
ordinary workman who works for wages
only, does not feel this love ; the professional
man whose profession is simply a source
of income, is little better ; but education
can be so directed that no man can really
enter into the spirit of the work, for which
this education has trained him, without

caring more for the work than for the profit, without an interest which is really love. The men who are to save the new civilization from business trickery, commercial cruelty, and selfish indulgence must feel this interest in the work they do; they must seek the best results because they love the best; they must do their work because they love it, not perhaps with all their heart and soul, but with the full strength of their intellectual capacity. This love for their work has characterized the best students and investigators in all ages. With the change which the manufacture of power has introduced, it should exist in every branch of work which deals with the utilization of the great sources of power in nature. The university will fail in its duty to the community if it does not inspire young men with a love for their work.

VII

THE educational system of our country is properly divided into four grades. It begins with the common-school system where children first begin to learn, and where they get the education which every man and woman ought to have; this much is required by law; without it a child grows up in ignorance. Then follows the secondary school, with a higher education which no one is compelled to take, but an education which is generally sought for by all who have any taste for study, and who do not have to work for wages as soon as they leave the common school. Next comes the college, whose duty is to round out the general education, and to equip the youth with that general knowledge and accomplishment which places him in the rank of educated men. The last is the professional school, with its special training for the vocation the student is to follow. The number who will enter each successive grade must always be less than those who take the grade below. The college course is omitted by many who go directly from secondary to professional schools.

Of the common school little need be said. It has become so thoroughly embodied in American institutions that the younger generation finds it hard to realize that there is any place where it does not exist. The secondary school is another thing, although it has often been accepted as a proper public charge. It includes both high and preparatory schools ; the scholars in both are of the same age, and the studies on parallel lines, though the high school is intended to give an education from which boys will enter business life, and the preparatory school trains for something which is to follow. The secondary school has one advantage over any other : its scholars are at the age in which the mind is most easily and most thoroughly trained ; the rudimentary education of the common school has given them some tools to work with ; they are not so old that mental habits are hard to form ; it takes the boy when he begins to know the difference between what he is taught and what he thinks ; while there may be no period in his school life the studies of which he will remember less, there is certainly no period in which the work has so much effect. The possible influence of a teacher in a secondary school is enormous ; the great principal who for fifty years was at the head of one of these schools in New England[1] exerted an indirect power over his whole country which few men ever had. Many a man whose name is now synonymous with greatness traced the awakening of his mind to this teacher.

The function of the college is different. The boy who has begun to think for himself, whose tastes and habits are partly

[1] Benjamin Abbot, LL. D.

formed, must here get the general training
which places him among educated men ; a
training which, going outside of particular
callings, provides that general equipment
of varied knowledge which fits him to as-
sociate with men of other occupations, and
at least to appreciate lines of work and study
which he does not follow himself.

What the college should avoid the professional school is bound to do ; the young
man comes here with his plan of life formed,
he asks for the special training which will
fit him to carry out his plan. The time for
general education has gone by, and the
duties of the professional school are specific.

In earlier times there was but a single
course of education from beginning to end,
and this was taken by comparatively few.
The working classes received no education ;
the ruling classes, who were generally soldiers, received but little ; women were given
none. Between the ruling and the working
classes was the clerical body, the only educated set, who formed the clerks and writers

for their more ignorant masters ; the Egyptian officer of to-day need not know how to read and write when he has secretaries to do it for him. But the demands of education have changed; every profession and occupation calls for educated men, and the education of to-day must fit them for this work, with all its variety. No single curriculum can meet the requirements of the new epoch ; the universal scholar can no longer exist; the man who tries to know every thing must spend his life in study alone, with no time left to make use of his accomplishments ; he may be admired for his great learning, but his education will be like some tariffs, continued long after its usefulness has gone. Every man who would contribute what he owes to the world of which he is a part, must accept his true position in his own department of study and of subsequent work ; he cannot know all, but he can console himself with the thought that he does not live alone, that he is one of a community, and that as a member of such community he has an interest and a part in what all the others are doing.

It is not long since few of the occupations which are included in civil engineering required any high order of preparation. The fact that some of the ablest engineers in the middle of the century had barely a common-school education is sometimes used as an argument against professional training. These old engineers did much; some of them were great men; they are entitled to honor and respect; they were the pioneers of the profession; but there is a great difference between the work of a pioneer and the work which follows. The pioneer must tread on unknown ground, he cannot be educated in those things which he must himself discover; judgment, sound sense, and daring distinguish the work of the pioneer; but the work of his followers, who are able to study what he could only explore, is better than his own. In the romantic town of Cuernavaca the Spanish conqueror built a church whose massive roughly buttressed walls still resist the thrust of the clumsy arched stone roof. On the plaza of the city of Mexico stands a great cathedral, built nearly two centuries later, which shows in every line of its groins and domes, its columns and its walls, the intelligence of its designer; structurally honest, it has little in common with the fashionable work by which our cities are scarred, but it has a high rank among the noble buildings of the world. The two churches typify the work of the pioneer and the work which should follow. An established profession has no longer pioneers. The men who to-day are to direct the great powers of nature to the use of man, who are to make the records of the future, must understand the laws by which they are to do this. The greatest engineer is not the man who knows the most, but the man who, when confronted with a new problem, can best grasp the novel subject, and whose judgment will most correctly approve or condemn his solution of it. This is the proper qualification for individual engineers, but even this is not enough for the profession.

Liberality means much the same in education and in professions as in anything else. It means freedom from the narrow boundaries of prejudice, and it recognizes that there is something beyond the contracted limits which bound the education or the profession of any one man. The real difference between a liberal education and a special education is that the first teaches the student to use his mind, and the second supplies him with information; the man who has only a special education may be no better than the uneducated man among surroundings for which he is not educated. The distinctive mark of a liberal profession is that its members shall know enough of matters outside their own professional work, to respect, to appreciate, and in a general way to understand the work of other men, whether those men belong to other active professions or to the various educated callings which are not given professional names.

The education which will keep civil engineering where it should be — in its rank among the liberal professions — must be broad enough for its disciples to understand and respect other lines of life ; it must make it a profession whose members will hold their own among other educated men, a profession which will have its proper share of writers and original thinkers, a profession through which men may rise to do the most influential work, and to take the highest places of trust and honor in the land.

There is a great difference between the work of the museum and that of a school, between the old and the new education, between the study of the past and the training for the future ; the education of different individuals must first divide into these two general directions. But there are several professions and a multitude of occupations based on the history of the past, and grounded on the old education of the museum ; and there are many vocations which are gradually being grouped into separate professions, and which are grounded on the scientific courses of the new education. There must be a division first between the museum and the school, between the new and the old education ; and there must be another division when each of these two great educations separates into many courses leading in different professional lines. The great educational question of the day is how to provide for these divisions of work ; it is a problem with which our universities are wrestling; it is a problem which is now thrown back on the preparatory schools, whose work is largely determined by the colleges to which they send their pupils ; it finds its full development in the courses of the professional schools, which must train for separate professions and specialties.

The separation must begin in the second-ary schools, but this is a beginning rather than more. Under the training of these schools, boys begin to show their natural aptitudes, and some of their studies first take a direction which may lead to the classic

shades of the museum or the more active
work of modern life. During the college
years the first separation must be completed;
the work of the museum and of the school
must diverge. A college course should be
framed with reference to the mental apti-
tude of the student and the general order
of life which he is likely to follow; while
there might be courses which would combine equal selections from the two, it must
generally be a course distinctly belonging
either to the old education or to the new;
but each should include enough of the other
to give the student that general knowledge
which every liberally educated man must
have, and to develop that respect for other
work which is the highest characteristic of
the cultivated man.

Changes have taken place in college
courses ; fifty years ago they were much the
same in all American colleges, and for every
student at each college; but even then they
covered a variety of subjects not found
in any other educational institution. The

studies were of the past rather than for the future; they were classical and historical, rather than scientific; the old education was generally the only one. A great variety of new studies have now been introduced, while the methods of conducting the old ones have changed. A paper warfare has gone on between science and the classics, and much is heard of classical and scientific courses ; if not carried too far this is all right. The subjects which now find their place in a college course are too many for any one to take them all, but a selection should include studies of very different kinds. The studies which are generally supposed to form a classical course, though no college course now consists exclusively of classical studies, are the ones best fitted to train the young man whose life is to be devoted to the work of the museum, and to those professions which are based on a history of the past; they are still the bulwarks of the old education. A scientific course, which ought in no college to consist entirely of science, is the true beginning of the new education, and the one which, in the order of other departments, would precede the professional work of the engineering school.

Different methods have been adopted to meet the increased variety of college studies. But few of the studies which are now taught at colleges can be taken by any one man. The simplest method has been to provide instruction in a large number of studies and let each student select what he would take himself, simply requiring that a sufficient aggregate amount of work be done ; while this method may sometimes result well it assumes a capacity of choice which few young men have ; few college students know, except in a general way, what studies they wish to pursue ; scarcely any college student is capable of making that judicious selection by which he will get a general training in the direction in which his mind leads him, with a sufficient admixture of other studies to meet the general requirements of a truly liberal education ; the desultory selection

which some boys have made has produced
college graduates the Latin of whose degrees is a strange travesty on the ignorance
of their minds. Another method, which requires special skill in the college government, but from which good results may
come, is to group studies in courses and require each student to select a course which,
arranged by older men than himself, may be
at least expected to have the proper admixture for good results. With colleges of moderate size this may be the best method.
When a college becomes very large there is
a method, which has been little tried, but
which is probably the best of all ; it is not
merely to divide the work into separate
courses, each being a wisely selected group
of studies, but to divide the college itself
into a number of smaller colleges, each having
but a single course, the number of colleges
corresponding to the number of courses ; this
scheme will involve duplicate instruction if
each college does all its own work, but it
has the great advantage of grouping to-

gether in one college the young men who
pursue each course of studies and who have
a single purpose ; it calls upon every student
to measure himself with his fellows in a
spirit of friendly rivalry, and it generates
that mutual esteem which comes when
young men of very different home training,
but who are now working for a common end,
are brought together in the intimate relations of daily life.

With the professional school the separation of each class of education into its final
subdivision is complete. The first division
comes gradually ; it takes years from its
first beginning in the secondary school to
the end of the college course ; the subdivision which follows comes all at once when
the young man is ready to devote all his
energies to the single pursuit of his profession.

The three older professions, which were
formerly the only ones recognized as liberal, are law, medicine, and divinity. Special professional schools exist for each of

these. The first and third belong to the
old education, but medicine deals with the
powers in nature for the use of man; it
differs from civil engineering principally in
that it deals with organic life and not with
inanimate power; its recent developments
have been rendered possible by the same
conditions that have developed the engineering profession; its place is with the
other branches of physical science in the
new education. The courses of these three
sets of professional schools are generally
based on the existence of a previous college
course, which it is at least hoped that students entering them will have taken. One
leading American university, which has been
at great pains to raise the standard of these
three departments, now limits admission to
its Law School to graduates of colleges, or
to persons qualified to enter the senior
class of its own college; it requires an
examination for admission to its Medical
School, but accepts a college degree as a
substitute for such examination; a candi-

date for a degree in its Divinity School must either be a college graduate or he must give evidence that he has received an education at least equal to what a college would give. These are now recognized as the requirements which should precede the special work of the professions which were formerly known as the only liberal professions. The requirements are right; specialists can be trained without these earlier studies; liberal professions need them. While it would neither be just nor wise to close the bar, the pulpit, or the hospital to men without college educations, it is well that at least one university should declare that it will give no professional degrees to men who are not qualified to make their professions liberal.

Similar requirements are seldom made as conditions for admission to schools of engineering. The same university which now insists on such rigid requirements for admission to its older professional schools, is satisfied with lower requirements for its

Scientific School than it asks for its undergraduate department; while it has raised the other professional schools to a strictly postgraduate rank, it has given its Scientific School a rank inferior to that of its classical college, though this Scientific School is its only provision for educating engineers.

The demands of the engineering profession are now at least equal to those of any other. The time required to master the special studies of the profession is certainly no less than the time needed for law or medicine or theology. The broad definition of civil engineering embraces all the branches of the engineering profession, and an institution which would train for civil engineering in its broadest sense must really comprise several professional schools. The requirements now demanded by any engineering specialty are enough to occupy the full time of the student during the years ordinarily allotted to professional schools. If the professional school of engineering can dispense with the previous training of the college, it means that the engineer needs less general education than members of other professions do ; it would almost seem to mean that civil engineering is not a liberal profession. Undoubtedly the work of the engineer is less dependent on a knowledge of the past than is that of the lawyer or the priest. An engineer can separate himself entirely from all work and study outside of close professional lines without impairing his strictly professional ability. The most skillful specialist may be entirely ignorant of everything but his own work. The man who knows nothing but his own profession may be the most useful man to his employer, even though he be employed in a position of great importance and liberal pay. From this point of view, the additional years which would be spent in a college may be considered wasted, and the boy who enters a technical school would be wise in saving the time which would be unnecessarily spent in a more varied education.

The real question, however, is not the
value of the college course to any individual
engineer, but the value of this preliminary
education to the profession as a whole. Education must not be organized on the selfish
basis of business or for personal ends. The
true education for a profession is the education which will make that profession accomplish the most for the world. Every professional man must have the education of the
common school, and of the secondary school.
Every professional man, be he engineer,
lawyer, or physician, may, if he wishes, omit
the college, though the omission closes
some avenues of later professional studies ;
the omission may not impair his professional efficiency, it may only deprive him of
the general education which will help him
to give his own profession a standing among
the other liberal professions. The omission
of the college course, thereby bringing the
professional schools of engineering immediately after the secondary schools, is right,
if all engineers are to confine themselves to

the strict limits of their work, if all engineers will be satisfied to be employed rather than to be employers, if engineers are to live by themselves, and measure the value of life by what they alone do. But the civil engineer ought to fill many positions which are now held by men of different training; a thorough education in the laws which govern the construction and the working of the tools which men are using must be the best qualification for the control and use of such tools. Among the most successful railroad managers of to-day are those who have been first trained as engineers, and have been disciplined by the painstaking care of physical management. The best, the most important, and the most honorable work is that which must be performed by comparatively few, by men who, passing beyond the strict lines of professional work, occupy those exceptional positions in which they control others, in which they direct the progress of their race.

Amid the plains and hills of India may

be seen great cities which were built or
which were abandoned by the orders and at
the pleasure of absolute rulers. We cannot
build cities in this way here; their location,
their growth, and their prosperity must depend on commercial developments, and on
laws which no one man can control. An
absolute monarch might create a university
at the location which he preferred and
with the full organization which he thought
best; we cannot do this here. The development of our educational institutions must,
like everything else we do, be gradual, and
be determined chiefly by circumstances
which we cannot ourselves control. The
most that we can do is to give some slight
guidance to the unseen powers which are
at work, and, so far as our own lives and
influence can reach, direct these powers in
the right way.

A great university, which would be complete in all departments, would be divided
on two planes, — on the planes of time and of
subjects. As the head of an educational sys-

tem, it would not be improper for it to exercise supervision and direction over the common and the secondary schools, but the actual maintenance of inferior departments is hardly a proper part of its work. There would be but one division in time, — into the colleges or undergraduate schools and the professional or postgraduate schools. The division among subjects would be more varied; there would be, first, the great division based on the studies of the museum and the studies for active life, and then there would be various subdivisions to meet specific aptitudes and duties. In the undergraduate schools the line between the old and the new education would not be absolutely drawn. There would be a number of college courses, or better a number of colleges, whose work would vary all the way from that which is almost entirely classical to that which is almost entirely scientific, from that which thoroughly represents the old education to that which thoroughly represents the new. In the postgraduate schools, the division among subjects would be absolute ; these schools must belong to one education or the other ; there would be a distinct division between, on the one hand, the graduate schools of the museum, including those for the older professions, which are based on study of the past, and, on the other hand, the schools which instruct young men to utilize the powers of nature for the good of men, whether in the various lines of engineering, or in the courses which, dealing with organic life, will train the physicians and biologists of the future.

Such a complete university does not now exist. The name university is generally accorded to any group of schools and colleges which forms a considerable part of the great whole. A group of undergraduate colleges may be termed a university. A group of colleges and graduate schools, whose work is generally that of the old education, may be termed a university ; some of the older and more prominent universities of America are of this kind. A group of colleges and graduate schools belonging entirely to the new education may be called a university; some of the newer universities of America are based on these lines.

There would seem to be great advantage in grouping together in one university the several undergraduate colleges whose courses vary from the extreme of the old to the extreme of the new education, in uniting them where there may be some acquaintance and some communication between the students of even those colleges whose courses differ most widely, where equal rank and merit may lead each to respect the other. Liberal professions, professions whose members would appreciate what they cannot perform, and admire what they can only partially understand, demand not only that there should be some admixture of science with the classics, and of the classics with science, but also that the young men who choose one should not be kept too far away from those who choose the other.

On the other hand little may be gained

by placing the schools for different professions under the wing of a common university management. Professional training is
too distinctive a matter, its lines are too
closely drawn, and the students are generally too old, to have much to do with those
who are following other pursuits. There
might be satisfaction and some advantage
in having schools for all professions grouped
in a single university ; but while the studies
of the college course can be pursued in the
quiet of a university town, away from the
active work of the present day, perhaps
better than anywhere else, professional
education should not be separated too far
from the lines of work and the sympathies
of the profession itself ; there would be few
locations equally favorable for schools of
every profession. The associations of a college undergraduate should be with others
of the same age and educational rank ; the
associations of professional students should
be with older men of their own profession,
rather than with students of other profes-

sional schools. While the best college work can probably be done in those colleges which, with widely different courses, are grouped into a single university, the best professional instruction may be given by professional schools located where the students can have some intercourse with the older men of the professions they are to follow, and can at least see something of actual work of the kind they hope later to do themselves. There must be opportunities for interchange of ideas between professors and workers, between students and their future employers, between the young men and the old men whose places they expect to take.

VIII

CONCLUSION

WE are all familiar with the story of the prophet who, after trials and tribulations of strange and terrible character, went to a city and preached that its destruction was close at hand. This volume is not prepared in the spirit of prophecy, nor are the conditions of the new epoch like those which attended this tremendous preaching. Still in many ways the new epoch must open as an era of destruction. It must from its very nature destroy many of the conditions which give most interest to the history of the past, and many of the traditions which people hold most dear. It will put an end, once for all, to savage and barbarous races, who must either be elevated to the life of their more civilized contemporaries or must vanish from existence. It must destroy

ignorance, as the entire world will be educated, and one of the greatest dangers
must come from this very source, when the
number of half-educated people is greatest,
when the world is full of people who do
not know enough to recognize their limita-
tions, but know too much to follow loyally
the direction of better qualified leaders.
With the disappearance of ignorance will
come the destruction of superstitions which
have exercised such a momentous influence
in the past, perhaps generally for evil, but
in many instances for good. There must
be a great destruction, both in the physical
and in the intellectual world, of old buildings, old boundaries, and old monuments,
and furthermore of customs and ideas, systems of thought and methods of education.
There must be changes in governments,
which means destruction of old constitu-
tions, destruction of races as they merge
into one another, destruction of languages,
in the gradual disappearance of all but a
few of the most important, while even these

may in time give place to one that is universal. It is not worth while to consider how this destruction will occur, nor how much time it will require for its completion. The important fact is that destruction will come, not because the things which are destroyed are in themselves bad, but because however good and useful they may have been in the past, they are not adapted to fulfill the requirements of the new epoch.

Our principal thought must not be of the destruction but of the new development which makes that destruction necessary. The destruction is not something to be feared or avoided, it is inevitable. But destruction is always attended with danger; some time may elapse after the old has gone before the new is established in its place. If any warning is to be given, this is where it should come. Education must adapt itself to the new demands ; professional work must be extended, — the work of the men who receive salaries and fees for what they know and can do, rather than those who, on the one hand, take the chances of profits on their own ventures and speculations, and, those who, on the other hand, receive wages for physical rather than mental labor. There will be many other changes, the general nature of a few of which has already been outlined. The danger is that the destructive changes will come too fast, and the developments which are to take their place not fast enough. The trouble will lie in the possible gap between the two. The next two or three centuries may have periods of war, insurrection, and other trials, which it would be well if the world could avoid.

When the period of change is over it would seem as if the final conditions of human civilization would have been reached. It is proper to say that it seems, for no one can tell what new capacity may at some future time change the conditions of life again. But when the development of the new epoch has become universal over the whole globe, so that all have the same tools to work with, and if necessary to defend and fight with, so that each land has the advantage of what all the others know; when the possibilities of adventure are removed; when the opportunities for speculation no longer exist; it would seem as if mankind must settle down to a long period of rest. In many parts of our country, towns may be found which, after a few years of rapid, energetic development, seem to have attained their growth, and improvement seems to stop; but in reality they simply cease to increase rapidly in population; better houses are built, the people are surrounded with more comforts, and though the adventurous pioneers may be gone, the people who remain have settled down to the quiet satisfaction of a comfortable life. Similar changes have come in national life where nations became so consolidated or so enlarged that there was no further occasion for war, and their communication with other nations virtually ceased. Such was the history of Japan; such seems to have been the history of China. They have

prospered and continued in a morbid, self-
satisfied existence, until the manufacture of
power opened the new epoch in the western
world and they suddenly became aware that
other and stronger nations than themselves
existed. A like condition will occur when
the new epoch is fully developed, but with
one important difference; it will not be the
condition of a town nor of a nation but of
the whole earth, with nothing to change it
unless communication should be opened
with another planet, a possibility perhaps,
but one on which we need waste no thought.

It seems likely that material developments will come to a gradual pause, that
the stimulus will be removed, and that the
densely populated earth may continue for
centuries with comparatively little change;
that then an immense population will live
comfortably and happily, and the qualities
which make the good citizen and the contented man will be more in demand than
those which make leaders in such periods
as we are familiar with. It may be a time

when every one will understand the comfort and peace of mind which attend the adaptation of personal feeling to the general conditions which surround him. However satisfactory this condition may be, whatever its delights, and whatever the excellence and happiness which may be in store for the inhabitants of the thirtieth century, they are not for us. We have, however, one privilege which will not belong to them. In all history and in all periods of the world, the honors are held to belong, not to those who enjoy the results, but to those who have made these results possible. Our generation has the privilege of doing its full share in bringing forth the great changes which are ushering in the new epoch.

TECHNOLOGY AND SOCIETY

An Arno Press Collection

Ardrey, R[obert] L. **American Agricultural Implements.** In two parts. 1894

Arnold, Horace Lucien and Fay Leone Faurote. **Ford Methods and the Ford Shops.** 1915

Baron, Stanley [Wade]. **Brewed in America:** A History of Beer and Ale in the United States. 1962

Bathe, Greville and Dorothy. **Oliver Evans:** A Chronicle of Early American Engineering. 1935

Bendure, Zelma and Gladys Pfeiffer. **America's Fabrics:** Origin and History, Manufacture, Characteristics and Uses. 1946

Bichowsky, F. Russell. **Industrial Research.** 1942

Bigelow, Jacob. **The Useful Arts:** Considered in Connexion with the Applications of Science. 1840. Two volumes in one

Birkmire, William H. **Skeleton Construction in Buildings.** 1894

Boyd, T[homas] A[lvin]. **Professional Amateur:** The Biography of Charles Franklin Kettering. 1957

Bright, Arthur A[aron], Jr. **The Electric-Lamp Industry:** Technological Change and Economic Development from 1800 to 1947. 1949

Bruce, Alfred and Harold Sandbank. **The History of Prefabrication.** 1943

Carr, Charles C[arl]. **Alcoa, An American Enterprise.** 1952

Cooley, Mortimer E. **Scientific Blacksmith.** 1947

Davis, Charles Thomas. **The Manufacture of Paper.** 1886

Deane, Samuel. **The New-England Farmer,** or Georgical Dictionary. 1822

Dyer, Henry. **The Evolution of Industry.** 1895

Epstein, Ralph C. **The Automobile Industry:** Its Economic and Commercial Development. 1928

Ericsson, Henry. **Sixty Years a Builder:** The Autobiography of Henry Ericsson. 1942

Evans, Oliver. **The Young Mill-Wright and Miller's Guide.** 1850

Ewbank, Thomas. **A Descriptive and Historical Account of Hydraulic and Other Machines for Raising Water,** Ancient and Modern. 1842

Field, Henry M. **The Story of the Atlantic Telegraph.** 1893

Fleming, A. P. M. **Industrial Research in the United States of America.** 1917

Van Gelder, Arthur Pine and Hugo Schlatter. **History of the Explosives Industry in America.** 1927

Hall, Courtney Robert. **History of American Industrial Science.** 1954

Hungerford, Edward. **The Story of Public Utilities.** 1928

Hungerford, Edward. **The Story of the Baltimore and Ohio Railroad, 1827-1927.** 1928

Husband, Joseph. **The Story of the Pullman Car.** 1917

Ingels, Margaret. **Willis Haviland Carrier, Father of Air Conditioning.** 1952

Kingsbury, J[ohn] E. **The Telephone and Telephone Exchanges:** Their Invention and Development. 1915

Labatut, Jean and Wheaton J. Lane, eds. **Highways in Our National Life:** A Symposium. 1950

Lathrop, William G[ilbert]. **The Brass Industry in the United States.** 1926

Lesley, Robert W., John B. Lober and George S. Bartlett. **History of the Portland Cement Industry in the United States.** 1924

Marcosson, Isaac F. **Wherever Men Trade:** The Romance of the Cash Register. 1945

Miles, Henry A[dolphus]. **Lowell, As It Was, and As It Is.** 1845

Morison, George S. **The New Epoch:** As Developed by the Manufacture of Power. 1903

Olmsted, Denison. **Memoir of Eli Whitney, Esq.** 1846

Passer, Harold C. **The Electrical Manufacturers, 1875-1900.** 1953

Prescott, George B[artlett]. **Bell's Electric Speaking Telephone.** 1884

Prout, Henry G. **A Life of George Westinghouse.** 1921

Randall, Frank A. **History of the Development of Building Construction in Chicago.** 1949

Riley, John J. **A History of the American Soft Drink Industry:** Bottled Carbonated Beverages, 1807-1957. 1958

Salem, F[rederick] W[illiam]. **Beer, Its History and Its Economic Value as a National Beverage.** 1880

Smith, Edgar F. **Chemistry in America.** 1914

Steinman, D[avid] B[arnard]. **The Builders of the Bridge:** The Story of John Roebling and His Son. 1950

Taylor, F[rank] Sherwood. **A History of Industrial Chemistry.** 1957

Technological Trends and National Policy, Including the Social Implications of New Inventions. Report of the Subcommittee on Technology to the National Resources Committee. 1937

Thompson, John S. **History of Composing Machines.** 1904

Thompson, Robert Luther. **Wiring a Continent:** The History of the Telegraph Industry in the United States, 1832-1866. 1947

Tilley, Nannie May. **The Bright-Tobacco Industry, 1860-1929.** 1948

Tooker, Elva. **Nathan Trotter:** Philadelphia Merchant, 1787-1853. 1955

Turck, J. A. V. **Origin of Modern Calculating Machines.** 1921

Tyler, David Budlong. **Steam Conquers the Atlantic.** 1939

Wheeler, Gervase. **Homes for the People,** In Suburb and Country. 1855